CANTERBURY PIECES

LECTOR HOUSE PUBLIC DOMAIN WORKS

HAPPY READING!

CANTERBURY PIECES

SAMUEL BUTLER,
R. A. STREATFEILD

ISBN: 978-93-5614-620-4

Published: 1914

CANTERBURY PIECES

BY

SAMUEL BUTLER

Author of "Erewhon," "The Way of All Flesh," etc.

EDITED BY
R. A. STREATFEILD

1914

CONTENTS

Page

I. Darwin on the Origin of Species 1

- A Dialogue. 4
- Barrel-Organs 9
- Darwin on Species: From the Press, 21 February, 1863. 10
- Darwin on Species: From the Press, March 14th, 1863. 12
- Darwin on Species: From the Press, 18 March, 1863. 13
- Darwin on Species: From the Press, April 11th, 1863.. 14
- Darwin on Species: From the Press, 22nd June, 1863.. 15

II. Darwin among the Machines. 17

III. Lucubratio Ebria. 21

IV. A Note on "The Tempest" Act III, Scene I. 25

V. The English Cricketers 27

DARWIN ON THE ORIGIN OF SPECIES

PREFATORY NOTE

As the following dialogue embodies the earliest fruits of Butler's study of the works of Charles Darwin, with whose name his own was destined in later years to be so closely connected, and thus possesses an interest apart from its intrinsic merit, a few words as to the circumstances in which it was published will not be out of place.

Butler arrived in New Zealand in October, 1859, and about the same time Charles Darwin's Origin of Species *was published. Shortly afterwards the book came into Butler's hands. He seems to have read it carefully, and meditated upon it. The result of his meditations took the shape of the following dialogue, which was published on 20 December, 1862, in the* Press *which had been started in the town of Christ Church in May, 1861. The dialogue did not by any means pass unnoticed. On the 17th of January, 1863, a leading article (of course unsigned) appeared in the* Press, *under the title "Barrel-Organs," discussing Darwin's theories, and incidentally referring to Butler's dialogue. A reply to this article, signed A.M., appeared on the 21st of February, and the correspondence was continued until the 22nd of June, 1863. The dialogue itself, which was unearthed from the early files of the* Press, *mainly owing to the exertions of Mr. Henry Festing Jones, was reprinted, together with the correspondence that followed its publication, in the* Press *of June 8 and 15, 1912. Soon after the original appearance of Butler's dialogue a copy of it fell into the hands of Charles Darwin, possibly sent to him by a friend in New Zealand. Darwin was sufficiently struck by it to forward it to the editor of some magazine, which has not been identified, with the following letter:—*

Down, Bromley, Kent, S.E.
March 24 [1863].

(Private).

Mr. Darwin takes the liberty to send by this post to the Editor a New Zealand newspaper for the very improbable chance of the Editor having some spare space to reprint a Dialogue on Species. This Dialogue, written by some [sic] quite unknown to Mr. Darwin, is remarkable from its spirit and from giving so clear and accurate a view of Mr. D. [sic] theory. It is also remarkable from being published in a colony exactly 12 years old, in which it might have [sic] thought only material interests would have been regarded.

The autograph of this letter was purchased from Mr. Tregaskis by Mr. Festing Jones, and subsequently presented by him to the Museum at Christ Church. The letter cannot be dated with certainty, but since Butler's dialogue was published in December, 1862, and it is at least probable that the copy of the PRESS *which contained it was sent to Darwin shortly after it appeared, we may conclude with tolerable certainty that the letter was written in March, 1863. Further light is thrown on the controversy by a correspondence which took place between Butler and Darwin in 1865, shortly after Butler's return to England. During that year Butler had published a pamphlet entitled* THE EVIDENCE FOR THE RESURRECTION OF JESUS CHRIST AS GIVEN BY THE FOUR EVANGELISTS CRITICALLY EXAMINED, *of which he afterwards incorporated the substance into* THE FAIR HAVEN. *Butler sent a copy of this pamphlet to Darwin, and in due course received the following reply:—*

Down, Bromley, Kent.
September 30 [1865].

My dear Sir,—I am much obliged to you for so kindly sending me your Evidences, etc. We have read it with much interest. It seems to me written with much force, vigour, and clearness; and the main argument to me is quite new. I particularly agree with all you say in your preface.

I do not know whether you intend to return to New Zealand, and, if you are inclined to write, I should much like to know what your future plans are.

My health has been so bad during the last five months that I have been confined to my bedroom. Had it been otherwise I would have asked you if you could have spared the time to have paid us a visit; but this at present is impossible, and I fear will be so for some time.

With my best thanks for your present,

I remain,

My dear Sir,

Yours very faithfully,
Charles Darwin.

To this letter Butler replied as follows:—

15 Clifford's Inn, E.C.
October 1st, 1865.

Dear Sir,—I knew you were ill and I never meant to give you the fatigue of writing to me. Please do not trouble yourself to do so again. As you kindly ask my plans I may say that, though I very probably may return to New Zealand in three or four years, I have no intention of doing so before that time. My study is art, and anything else I may indulge in is only by-play; it may cause you some little wonder that at my age I should have started as an art student, and I may perhaps be permitted to explain that this was always my wish for years, that I had begun six years ago, as soon as ever I found that I could not conscientiously take orders; my father so strongly disapproved of the idea that I gave it up and went out to New Zealand, stayed there for five years, worked like a common servant, though on a run of my own, and sold out little more than a year ago, thinking that prices were going to fall—which they have since done. Being then rather at a loss what to do and my capital being all locked up, I took the opportunity to return to my old plan, and have been studying for the last ten years unremittingly. I hope that in three or four years more I shall be able to go on very well by myself, and then I may go back to New Zealand or no as circumstances shall seem to render advisable. I must apologise for so much detail, but hardly knew how to explain myself without it.

I always delighted in your ORIGIN OF SPECIES *as soon as I saw it out in New Zealand—not as knowing anything whatsoever of natural history, but it enters into so many deeply interesting questions, or rather it suggests so many, that it thoroughly fascinated me. I therefore feel all the greater pleasure that my pamphlet should please you, however full of errors.*

The first dialogue on the ORIGIN *which I wrote in the* PRESS *called forth a contemptuous rejoinder from (I believe) the Bishop of Wellington—(please do not mention the name, though I think that at this distance of space and time I might mention it to yourself) I answered it with the enclosed, which may amuse you. I assumed another character because my dialogue was in my hearing very severely criticised by two or three whose opinion I thought worth having, and I deferred to their judgment in my next. I do not think I should do so now. I fear you will be shocked at an appeal to the periodicals mentioned in my letter, but they form a very staple article of bush diet, and we used to get a good deal of superficial knowledge out of them. I feared to go in too heavy on the side of the* ORIGIN, *because I thought that, having said my say as well as I could, I had better now take a less impassioned tone; but I was really exceedingly angry.*

Please do not trouble yourself to answer this, and believe me,

Yours most sincerely,
S. Butler.

This elicited a second letter from Darwin:—

Down, Bromley, Kent.
October 6.

My dear Sir,—I thank you sincerely for your kind and frank letter, which has interested me greatly. What a singular and varied career you have already run. Did you keep any journal or notes in New Zealand? For it strikes me that with your rare powers of writing you might make a very interesting work descriptive of a colonist's life in New Zealand.

I return your printed letter, which you might like to keep. It has amused me, especially the part in which you criticise yourself. To appreciate the letter fully I ought to have read the bishop's letter, which seems to have been very rich.

You tell me not to answer your note, but I could not resist the wish to thank you for your letter.

With every good wish, believe me, my dear Sir,

Yours sincerely,
Ch. Darwin.

It is curious that in this correspondence Darwin makes no reference to the fact that he had already had in his possession a copy of Butler's dialogue and had endeavoured to induce the editor of an English periodical to reprint it. It is possible that we have not here the whole of the correspondence which passed between Darwin and Butler at this period, and this theory is supported by the fact that Butler seems to take for granted that Darwin knew all about the appearance of the original dialogue on the Origin of Species *in the* Press. *Enough, however, has been given to explain the correspondence which the publication of the dialogue occasioned. I do not know what authority Butler had for supposing that Charles John Abraham, Bishop of Wellington, was the author of the article entitled "Barrel-Organs," and the "Savoyard" of the subsequent controversy. However, at that time Butler was deep in the counsels of the* Press, *and he may have received private information on the subject. Butler's own reappearance over the initials A.M. is sufficiently explained in his letter to Darwin.*

It is worth observing that Butler appears in the dialogue and ensuing correspondence in a character very different from that which he was later to assume. Here we have him as an ardent supporter of Charles Darwin, and adopting a contemptuous tone with regard to the claims of Erasmus Darwin to have sown the seed which was afterwards raised to maturity by his grandson. It would be interesting to know if it was this correspondence that first turned Butler's attention seriously to the works of the older evolutionists and ultimately led to the production of Evolution, Old and New, *in which the indebtedness of Charles Darwin to Erasmus Darwin, Buffon and Lamarck is demonstrated with such compelling force.*

A DIALOGUE

[From the *Press*, 20 December, 1862.]

F. So you have finished Darwin? Well, how did you like him?

C. You cannot expect me to like him. He is so hard and logical, and he treats his subject with such an intensity of dry reasoning without giving himself the loose rein for a single moment from one end of the book to the other, that I must confess I have found it a great effort to read him through.

F. But I fancy that, if you are to be candid, you will admit that the fault lies rather with yourself than with the book. Your knowledge of natural history is so superficial that you are constantly baffled by terms of which you do not understand the meaning, and in which you consequently lose all interest. I admit, however, that the book is hard and laborious reading; and, moreover, that the writer appears to have predetermined from the commencement to reject all ornament, and simply to argue from beginning to end, from point to point, till he conceived that he had made his case sufficiently clear.

C. I agree with you, and I do not like his book partly on that very account. He seems to have no eye but for the single point at which he is aiming.

F. But is not that a great virtue in a writer?

C. A great virtue, but a cold and hard one.

F. In my opinion it is a grave and wise one. Moreover, I conceive that the judicial calmness which so strongly characterises the whole book, the absence of all passion, the air of extreme and anxious caution which pervades it throughout, are rather the result of training and artificially acquired self-restraint than symptoms of a cold and unimpassioned nature; at any rate, whether the lawyer-like faculty of swearing both sides of a question and attaching the full value to both is acquired or natural in Darwin's case, you will admit that such a habit of mind is essential for any really valuable and scientific investigation.

C. I admit it. Science is all head—she has no heart at all.

F. You are right. But a man of science may be a man of other things besides science, and though he may have, and ought to have no heart during a scientific investigation, yet when he has once come to a conclusion he may be hearty enough in support of it, and in his other capacities may be of as warm a temperament as even you can desire.

C. I tell you I do not like the book.

F. May I catechise you a little upon it?

C. To your heart's content.

F. Firstly, then, I will ask you what is the one great impression that you have derived from reading it; or, rather, what do you think to be the main impression that Darwin wanted you to derive?

C. Why, I should say some such thing as the following—that men are descended from monkeys, and monkeys from something else, and so on back to dogs and horses and hedge-sparrows and pigeons and cinipedes (what is a cinipede?) and cheesemites, and then through the plants down to duckweed.

F. You express the prevalent idea concerning the book, which as you express it appears nonsensical enough.

C. How, then, should you express it yourself?

F. Hand me the book and I will read it to you through from beginning to end, for to express it more briefly than Darwin himself has done is almost impossible.

C. That is nonsense; as you asked me what impression I derived from the book, so now I ask you, and I charge you to answer me.

F. Well, I assent to the justice of your demand, but I shall comply with it by requiring your assent to a few principal statements deducible from the work.

C. So be it.

F. You will grant then, firstly, that all plants and animals increase very rapidly, and that unless they were in some manner checked, the world would soon be overstocked. Take cats, for instance; see with what rapidity they breed on the different runs in this province where there is little or nothing to check them; or even take the more slowly breeding sheep, and see how soon 500 ewes become 5000 sheep under favourable circumstances. Suppose this sort of thing to go on for a hundred million years or so, and where would be the standing room for all the different plants and animals that would be now existing, did they not materially check each other's increase, or were they not liable in some way to be checked by other causes? Remember the quail; how plentiful they were until the cats came with the settlers from Europe. Why were they so abundant? Simply because they had plenty to eat, and could get sufficient shelter from the hawks to multiply freely. The cats came, and tussocks stood the poor little creatures in but poor stead. The cats increased and multiplied because they had plenty of food and no natural enemy to check them. Let them wait a year or two, till they have materially reduced the larks also, as they have long since reduced the quail, and let them have to depend solely upon occasional dead lambs and sheep, and they will find a certain rather formidable natural enemy called Famine rise slowly but inexorably against them and slaughter them wholesale. The first proposition then to which I demand your assent is that all plants and animals tend to increase in a high geometrical ratio; that they all endeavour to get that which is necessary for their own welfare; that, as unfortunately there are conflicting interests in Nature, collisions constantly occur between different animals and plants, whereby the rate of increase of each species is very materially checked. Do you admit this?

C. Of course; it is obvious.

F. You admit then that there is in Nature a perpetual warfare of plant, of bird, of beast, of fish, of reptile; that each is striving selfishly for its own advantage, and will get what it wants if it can.

C. If what?

F. If it can. How comes it then that sometimes it cannot? Simply because all are not of equal strength, and the weaker must go to the wall.

C. You seem to gloat over your devilish statement.

F. Gloat or no gloat, is it true or no? I am not one of those

> "Who would unnaturally better Nature
> By making out that that which is, is not."

If the law of Nature is "struggle," it is better to look the matter in the face and adapt yourself to the conditions of your existence. Nature will not bow to you, neither will you mend matters by patting her on the back and telling her that she is not so black as she is painted. My dear fellow, my dear sentimental friend, do you eat roast beef or roast mutton?

C. Drop that chaff and go back to the matter in hand.

F. To continue then with the cats. Famine comes and tests them, so to speak; the weaker, the less active, the less cunning, and the less enduring cats get killed off, and only the strongest and smartest cats survive; there will be no favouritism shown to animals in a state of Nature; they will be weighed in the balance, and the weight of a hair will sometimes decide whether they shall be found wanting or no. This being the case, the cats having been thus naturally culled and the stronger having been preserved, there will be a gradual tendency to improve manifested among the cats, even as among our own mobs of sheep careful culling tends to improve the flock.

C. This, too, is obvious.

F. Extend this to all animals and plants, and the same thing will hold good concerning them all. I shall now change the ground and demand assent to another statement. You know that though the offspring of all plants and animals is in the main like the parent, yet that in almost every instance slight deviations occur, and that sometimes there is even considerable divergence from the parent type. It must also be admitted that these slight variations are often, or at least sometimes, capable of being perpetuated by inheritance. Indeed, it is only in consequence of this fact that our sheep and cattle have been capable of so much improvement.

C. I admit this.

F. Then the whole matter lies in a nutshell. Suppose that hundreds of millions of years ago there existed upon this earth a single primordial form of the very lowest life, or suppose that three or four such primordial forms existed. Change of climate, of food, of any of the circumstances which surrounded any member of this first and lowest class of life would tend to alter it in some slight manner, and the alteration would have a tendency to perpetuate itself by inheritance. Many failures would doubtless occur, but with the lapse of time slight deviations would undoubtedly become permanent and inheritable, those alone being perpetuated which were beneficial to individuals in whom they appeared. Repeat the process with each deviation and we shall again obtain divergences (in the course of ages) differing more strongly from the ancestral form, and again those that enable their possessor to struggle for existence most efficiently will be preserved. Repeat this process for millions and millions of years, and, as it is impossible to assign any limit to variability, it would seem as though the present diversities of species must certainly have come about sooner or later, and that other divergences will continue to come about to the end of time. The great agent in this development of life has been competition. This has culled species after species, and secured that those alone should survive which were best fitted for the conditions by which they found themselves surrounded. Endeavour to take a bird's-eye view of the whole matter. See battle after battle, first in one part of the world, then in another, sometimes raging more fiercely and sometimes less; even as in human affairs war has always existed in some part of the world from the earliest known periods, and probably always will exist. While a species is conquering in one part of the world it is being subdued in another, and while its conquerors are indulging in their triumph down comes the fiat for their being culled and drafted out, some to life and some to death, and so forth *ad infinitum*.

C. It is very horrid.

F. No more horrid than that you should eat roast mutton or boiled beef.

C. But it is utterly subversive of Christianity; for if this theory is true the fall of man is entirely fabulous; and if the fall, then the redemption, these two being inseparably bound together.

F. My dear friend, there I am not bound to follow you. I believe in Christianity, and I believe in Darwin. The two appear irreconcilable. My answer to those who accuse me of inconsistency is, that both being undoubtedly true, the one must be reconcilable with the other, and that the impossibility of reconciling them must be only apparent and temporary, not real. The reconciliation will never be effected by planing a little off the one and a little off the other and then gluing them together with glue. People will not stand this sort of dealing, and the rejection of the one truth or of the other is sure to follow upon any such attempt being persisted in. The true course is to use the freest candour in the acknowledgment of the difficulty; to estimate precisely its real value, and obtain a correct knowledge of its precise form. Then and then only is there a chance of any satisfactory result being obtained. For unless the exact nature of the difficulty be known first, who can attempt to remove it? Let me re-state the matter once again. All animals and plants in a state of Nature are undergoing constant competition for the necessaries of life. Those that can hold their ground hold it; those that cannot hold it are destroyed. But as it also happens that slight changes of food, of habit, of climate, of circumjacent accident, and so forth, produce a slight tendency to vary in the offspring of any plant or animal, it follows that among these slight variations some may be favourable to the individual in whom they appear, and may place him in a better position than his fellows as regards the enemies with whom his interests come into collision. In this case he will have a better chance of surviving than his fellows; he will thus stand also a better chance of continuing the species, and in his offspring his own slight divergence from the parent type will be apt to appear. However slight the divergence, if it be beneficial to the individual it is likely to preserve the individual and to reappear in his offspring, and this process may be repeated *ad infinitum*. Once grant these two things, and the rest is a mere matter of time and degree. That the immense differences between the camel and the pig should have come about in six thousand years is not believable; but in six hundred million years it is not incredible, more especially when we consider that by the assistance of geology a very perfect chain has been formed between the two. Let this instance suffice. Once grant the principles, once grant that competition is a great power in Nature, and that changes of circumstances and habits produce a tendency to variation in the offspring (no matter how slight such variation may be), and unless you can define the possible limit of such variation during an infinite series of generations, unless you can show that there is a limit, and that Darwin's theory over-steps it, you have no right to reject his conclusions. As for the objections to the theory, Darwin has treated them with admirable candour, and our time is too brief to enter into them here. My recommendation to you is that you should read the book again.

C. Thank you, but for my own part I confess to caring very little whether my millionth ancestor was a gorilla or no; and as Darwin's book does not please me, I shall not trouble myself further about the matter.

BARREL-ORGANS

[From the *Press*, 17 January, 1863.]

Dugald Stewart in his *Dissertation on the Progress of Metaphysics* says: "On reflecting on the repeated reproduction of ancient paradoxes by modern authors one is almost tempted to suppose that human invention is limited, like a barrel-organ, to a specific number of tunes."

It would be a very amusing and instructive task for a man of reading and reflection to note down the instances he meets with of these old tunes coming up again and again in regular succession with hardly any change of note, and with all the old hitches and involuntary squeaks that the barrel-organ had played in days gone by. It is most amusing to see the old quotations repeated year after year and volume after volume, till at last some more careful enquirer turns to the passage referred to and finds that they have all been taken in and have followed the lead of the first daring inventor of the mis-statement. Hallam has had the courage, in the supplement to his *History of the Middle Ages*, p. 398, to acknowledge an error of this sort that he has been led into.

But the particular instance of barrel-organism that is present to our minds just now is the Darwinian theory of the development of species by natural selection, of which we hear so much. This is nothing new, but a *réchauffée* of the old story that his namesake, Dr. Darwin, served up in the end of the last century to Priestley and his admirers, and Lord Monboddo had cooked in the beginning of the same century. We have all heard of his theory that man was developed directly from the monkey, and that we all lost our tails by sitting too much upon that appendage.

We learn from that same great and cautious writer Hallam in his *History of Literature* that there are traces of this theory and of other popular theories of the present day in the works of Giordano Bruno, the Neapolitan who was burnt at Rome by the Inquisition in 1600. It is curious to read the titles of his works and to think of Dugald Stewart's remark about barrel-organs. For instance he wrote on "The Plurality of Worlds," and on the universal "Monad," a name familiar enough to the readers of *Vestiges of Creation*. He was a Pantheist, and, as Hallam says, borrowed all his theories from the eclectic philosophers, from Plotinus and the Neo-Platonists, and ultimately they were no doubt of Oriental origin. This is just what has been shown again and again to be the history of German Pantheism; it is a mere barrel-organ repetition of the Brahman metaphysics found in Hindu cosmogonies. Bruno's theory regarding development of species was in Hallam's words: "There is nothing so small or so unimportant but that a portion of spirit dwells in it; and this spiritual substance requires a proper subject to become a plant or an animal"; and Hallam in a note on this passage observes how the modern theories of equivocal generation correspond with Bruno's.

No doubt Hallam is right in saying that they are all of Oriental origin. Pythagoras borrowed from thence his kindred theory of the metempsychosis, or transmigration of souls. But he was more consistent than modern philosophers; he recognised a downward development as well as an upward, and made morality and immorality the crisis and turning-point of change—a bold lion developed into a brave warrior, a drunken sot developed into a wallowing pig, and Darwin's slave-making ants, p. 219, would have been formerly Virginian cotton and tobacco growers.

Perhaps Prometheus was the first Darwin of antiquity, for he is said to have begun his creation from below, and after passing from the invertebrate to the sub-vertebrate, from thence to the backbone, from the backbone to the mammalia, and from the mammalia to the manco-cerebral, he compounded man of each and all:—

> Fertur Prometheus addere principi
> Limo coactus particulam undique
> Desectam et insani leonis
> Vim stomacho apposuisse nostro.

One word more about barrel-organs. We have heard on the undoubted authority of ear and eyewitnesses, that in a neighbouring province there is a church where the psalms are sung to a barrel-organ, but unfortunately the psalm tunes come in the middle of the set, and the jigs and waltzes have to be played through before the psalm can start. Just so is it with Darwinism and all similar theories. All his fantasias, as we saw in a late article, are made to come round at last to religious questions, with which really and truly they have nothing to do, but were it not for their supposed effect upon religion, no one would waste his time in reading about the possibility of Polar bears swimming about and catching flies so long that they at last get the fins they wish for.

DARWIN ON SPECIES:
FROM THE PRESS, 21 FEBRUARY, 1863.

To the Editor of the *Press*.

Sir—In two of your numbers you have already taken notice of Darwin's theory of the origin of species; I would venture to trespass upon your space in order to criticise briefly both your notices.

The first is evidently the composition of a warm adherent of the theory in question; the writer overlooks all the real difficulties in the way of accepting it, and, caught by the obvious truth of much that Darwin says, has rushed to the conclusion that all is equally true. He writes with the tone of a partisan, of one deficient in scientific caution, and from the frequent repetition of the same ideas manifest in his dialogue one would be led to suspect that he was but little versed in habits of literary composition and philosophical argument. Yet he may fairly claim the merit of having written in earnest. He has treated a serious subject seriously according to his lights; and though his lights are not brilliant ones, yet he has apparently done his best to show the theory on which he is writing in its most favourable aspect. He is rash, evidently well satisfied with himself, very possibly mistaken, and just one of those persons who (without intending it) are more apt to mislead than to lead the few people that put their trust in them. A few will always follow them, for a strong faith is always more or less impressive upon persons who are too weak to have any definite and original faith of their own. The second writer, however, assumes a very different tone. His arguments to all practical intents and purposes run as follows:—

Old fallacies are constantly recurring. Therefore Darwin's theory is a fallacy.

They come again and again, like tunes in a barrel-organ. Therefore Darwin's theory is a fallacy.

Hallam made a mistake, and in his *History of the Middle Ages*, p. 398, he corrects himself. Therefore Darwin's theory is wrong.

Dr. Darwin in the last century said the same thing as his son or grandson says now—will the writer of the article refer to anything bearing on natural selection and the struggle for existence in Dr. Darwin's work?—and a foolish nobleman said something foolish about monkey's tails. Therefore Darwin's theory is wrong.

Giordano Bruno was burnt in the year 1600 A.D.; he was a Pantheist; therefore Darwin's theory is wrong.

And finally, as a clinching argument, in one of the neighbouring settlements there is a barrel-organ which plays its psalm tunes in the middle of its jigs and waltzes. After this all lingering doubts concerning the falsehood of Darwin's theory must be at an end, and any person of ordinary common sense must admit that the theory of development by natural selection is unwarranted by experience and reason.

The articles conclude with an implied statement that Darwin supposes the Polar bear to swim about catching flies for so long a period that at last it gets the fins it wishes for.

Now, however sceptical I may yet feel about the truth of all Darwin's theory, I cannot sit quietly by and see him misrepresented in such a scandalously slovenly manner. What Darwin does say is that sometimes diversified and changed habits may be observed in individuals of the same species; that is that there are eccentric animals just as there are eccentric men. He adduces a few instances and winds up by saying that "in North America the black bear was seen by Hearne swimming for hours with widely open mouth, thus catching—almost like a whale—insects in the water." This and nothing more. (See pp. 201 and 202.)

Because Darwin says that a bear of rather eccentric habits happened to be seen by Hearne swimming for hours and catching insects almost like a whale, your writer (with a carelessness hardly to be reprehended in sufficiently strong terms) asserts by implication that Darwin supposes the whale to be developed from the bear by the latter having had a strong desire to possess fins. This is disgraceful.

I can hardly be mistaken in supposing that I have quoted the passage your writer alludes to. Should I be in error, I trust he will give the reference to the place in which Darwin is guilty of the nonsense that is fathered upon him in your article.

It must be remembered that there have been few great inventions in physics or discoveries in science which have not been foreshadowed to a certain extent by speculators who were indeed mistaken, but were yet more or less on the right scent. Day is heralded by dawn, Apollo by Aurora, and thus it often happens that a real discovery may wear to the careless observer much the same appearance as an exploded fallacy, whereas in fact it is widely different. As much caution is due in the rejection of a theory as in the acceptation of it. The first of your writers is too hasty in accepting, the second in refusing even a candid examination.

Now, when the *Saturday Review*, the *Cornhill Magazine, Once a Week*, and *Macmillan's Magazine*, not to mention other periodicals, have either actually and completely as in the case of the first two, provisionally as in the last mentioned, given their adherence to the theory in question, it may be taken for granted that the arguments in its favour are sufficiently specious to have attracted the attention and approbation of a considerable number of well-educated men in England. Three months ago the theory of development by natural selection was openly supported by Professor Huxley before the British Association at Cambridge. I am not adducing Professor Huxley's advocacy as a proof that Darwin is right (indeed, Owen opposed him tooth and nail), but as a proof that there is sufficient to be said on Darwin's side to demand more respectful attention than your last writer has thought it worth while to give it. A theory which the British Association is discussing with great care in England is not to be set down by off-hand nicknames in Canterbury.

To those, however, who do feel an interest in the question, I would venture to give a word or two of advice. I would strongly deprecate forming a hurried opinion for or against the theory. Naturalists in Europe are canvassing the matter with the utmost diligence, and a few years must show whether they will accept the theory or no. It is plausible; that can be decided by no one. Whether it is true or no can be decided only among naturalists themselves. We are outsiders, and most of us must be content to sit on the stairs till the great men come forth and give us the benefit of their opinion.

I am, Sir,
Your obedient servant,
A. M.

DARWIN ON SPECIES:
FROM THE PRESS, MARCH 14TH, 1863.

To the Editor of the *Press*.

Sir—A correspondent signing himself "A. M." in the issue of February 21st says:—"Will the writer (of an article on barrel-organs) refer to anything bearing upon natural selection and the struggle for existence in Dr. Darwin's work?" This is one of the trade forms by which writers imply that there is no such passage, and yet leave a loophole if they are proved wrong. I will, however, furnish him with a passage from the notes of Darwin's *Botanic Garden*:—

"I am acquainted with a philosopher who, contemplating this subject, thinks it not impossible that the first insects were anthers or stigmas of flowers, which had by some means loosed themselves from their parent plant; and that many insects have gradually in long process of time been formed from these, some acquiring wings, others fins, and others claws, from their ceaseless efforts to procure their food or to secure themselves from injury. The anthers or stigmas are therefore separate beings."

This passage contains the germ of Mr. Charles Darwin's theory of the origin of species by natural selection:—

"Analogy would lead me to the belief that all animals and plants have descended from one prototype."

Here are a few specimens, his illustrations of the theory:—

"There seems to me no great difficulty in believing that natural selection has actually converted a swim-bladder into a lung or organ used exclusively for respiration." "A swim-bladder has apparently been converted into an air-breathing lung." "We must be cautious in concluding that a bat could not have been formed by natural selection from an animal which at first could only glide through the air." "I can see no insuperable difficulty in further believing it possible that the membrane-connected fingers and forearm of the galeopithecus might be greatly lengthened by natural selection, and this, as far as the organs of flight are concerned, would convert it into a bat." "The framework of bones being the same in the hand of a man, wing of a bat, fin of a porpoise, and leg of a horse, the same number of vertebræ forming the neck of the giraffe and of the elephant, and innumerable other such facts, at once explain themselves on the theory of descent with slow and slight successive modifications."

I do not mean to go through your correspondent's letter, otherwise "I could hardly reprehend in sufficiently strong terms" (and all that sort of thing) the perversion of what I said about Giordano Bruno. But "ex uno disce omnes"—I am, etc.,

"The Savoyard."

DARWIN ON SPECIES:
FROM THE PRESS, 18 MARCH, 1863.

To the Editor of the Press.

Sir—The "Savoyard" of last Saturday has shown that he has perused Darwin's *Botanic Garden* with greater attention than myself. I am obliged to him for his correction of my carelessness, and have not the smallest desire to make use of any loopholes to avoid being "proved wrong." Let, then, the "Savoyard's" assertion that Dr. Darwin had to a certain extent forestalled Mr. C. Darwin stand, and let my implied denial that in the older Darwin's works passages bearing on natural selection, or the struggle for existence, could be found, go for nought, or rather let it be set down against me.

What follows? Has the "Savoyard" (supposing him to be the author of the article on barrel-organs) adduced one particle of real argument the more to show that the real Darwin's theory is wrong?

The elder Darwin writes in a note that "he is acquainted with a philosopher who thinks it not impossible that the first insects were the anthers or stigmas of flowers, which by some means, etc. etc." This is mere speculation, not a definite theory, and though the passage above as quoted by the "Savoyard" certainly does contain the germ of Darwin's theory, what is it more than the crudest and most unshapen germ? And in what conceivable way does this discovery of the egg invalidate the excellence of the chicken?

Was there ever a great theory yet which was not more or less developed from previous speculations which were all to a certain extent wrong, and all ridiculed, perhaps not undeservedly, at the time of their appearance? There is a wide difference between a speculation and a theory. A speculation involves the notion of a man climbing into a lofty position, and descrying a somewhat remote object which he cannot fully make out. A theory implies that the theorist has looked long and steadfastly till he is clear in his own mind concerning the nature of the thing which he is beholding. I submit that the "Savoyard" has unfairly made use of the failure of certain speculations in order to show that a distinct theory is untenable.

Let it be granted that Darwin's theory has been foreshadowed by numerous previous writers. Grant the "Savoyard" his Giordano Bruno, and give full weight to the barrel-organ in a neighbouring settlement, I would still ask, has the theory of natural development of species ever been placed in anything approaching its present clear and connected form before the appearance of Mr. Darwin's book? Has it ever received the full attention of the scientific world as a duly organised theory, one presented in a tangible shape and demanding investigation, as the conclusion arrived at by a man of known scientific attainments after years of patient toil? The upshot of the barrel-organs article was to answer this question in the affirmative and to pooh-pooh all further discussion.

It would be mere presumption on my part either to attack or defend Darwin, but my indignation was roused at seeing him misrepresented and treated disdainfully. I would wish, too, that the "Savoyard" would have condescended to notice that little matter of the bear. I have searched my copy of Darwin again and again to find anything relating to the subject except what I have quoted in my previous letter.

I am, Sir, your obedient servant,

A. M.

DARWIN ON SPECIES:
FROM THE PRESS, APRIL 11TH, 1863.

To the Editor of the Press.

Sir—Your correspondent "A. M." is pertinacious on the subject of the bear being changed into a whale, which I said Darwin contemplated as not impossible. I did not take the trouble in any former letter to answer him on that point, as his language was so intemperate. He has modified his tone in his last letter, and really seems open to the conviction that he may be the "careless" writer after all; and so on reflection I have determined to give him the opportunity of doing me justice.

In his letter of February 21 he says: "I cannot sit by and see Darwin misrepresented in such a scandalously slovenly manner. What Darwin does say is 'that

SOMETIMES diversified and changed habits may be observed in individuals of the same species; that is, that there are certain eccentric animals as there are certain eccentric men. He adduces a few instances, and winds up by saying that in North America the black bear was seen by Hearne swimming for hours with widely open mouth, thus catching, ALMOST LIKE A WHALE, insects in the water.' THIS, AND NOTHING MORE, pp. 201, 202."

Then follows a passage about my carelessness, which (he says) is hardly to be reprehended in sufficiently strong terms, and he ends with saying: "This is disgraceful."

Now you may well suppose that I was a little puzzled at the seeming audacity of a writer who should adopt this style, when the words which follow his quotation from Darwin are (in the edition from which I quoted) as follows: "Even in so extreme a case as this, if the supply of insects were constant, and if better adapted competitors did not already exist in the country, I can see no difficulty in a race of bears being rendered by natural selection more and more aquatic in their structure and habits, with larger and larger mouths, till a creature was produced as monstrous as a whale."

Now this passage was a remarkable instance of the idea that I was illustrating in the article on "Barrel-organs," because Buffon in his *Histoire Naturelle* had conceived a theory of degeneracy (the exact converse of Darwin's theory of ascension) by which the bear might pass into a seal, and that into a whale. Trusting now to the fairness of "A. M." I leave to him to say whether he has quoted from the same edition as I have, and whether the additional words I have quoted are in his edition, and if so whether he has not been guilty of a great injustice to me; and if they are not in his edition, whether he has not been guilty of great haste and "carelessness" in taking for granted that I have acted in so "disgraceful" a manner.

I am, Sir, etc.,
"The Savoyard," or player
on Barrel-organs.

(The paragraph in question has been the occasion of much discussion. The only edition in our hands is the third, seventh thousand, which contains the paragraph as quoted by "A. M." We have heard that it is different in earlier editions, but have not been able to find one. The difference between "A. M." and "The Savoyard" is clearly one of different editions. Darwin appears to have been ashamed of the inconsequent inference suggested, and to have withdrawn it. — Ed. the *Press*.)

DARWIN ON SPECIES:
FROM THE PRESS, 22ND JUNE, 1863.

To the Editor of the *Press*.

Sir—I extract the following from an article in the *Saturday Review* of January 10, 1863, on the vertebrated animals of the Zoological Gardens.

"As regards the ducks, for example, inter-breeding goes on to a very great extent among nearly all the genera, which are well represented in the collection. We think it unfortunate that the details of these crosses have not hitherto been made public. The Zoological Society has existed about thirty-five years, and we imagine that evidence must have been accumulated almost enough to make or mar that part of Mr. Darwin's well-known argument which rests on what is known of the phenomena of hybridism. The present list reveals only one fact bearing on the subject, but that is a noteworthy one, for it completely overthrows the commonly accepted theory that the mixed offspring of different species are infertile *inter se*. At page 15 (of the list of vertebrated animals living in the gardens of the Zoological Society of London, Longman and Co., 1862) we find enumerated three examples of hybrids between two perfectly distinct species, and even, according to modern classification, between two distinct genera of ducks, for three or four generations. There can be little doubt that a series of researches in this branch of experimental physiology, which might be carried on at no great loss, would place zoologists in a far better position with regard to a subject which is one of the most interesting if not one of the most important in natural history."

I fear that both you and your readers will be dead sick of Darwin, but the above is worthy of notice. My compliments to the "Savoyard."

Your obedient servant,
May 17th.
A. M.

DARWIN AMONG THE MACHINES

"Darwin Among the Machines" originally appeared in the Christ Church PRESS, 13 *June, 1863. It was reprinted by Mr. Festing Jones in his edition of* THE NOTE-BOOKS OF SAMUEL BUTLER (*Fifield, London, 1912, Kennerley, New York), with a prefatory note pointing out its connection with the genesis of* EREWHON, *to which readers desirous of further information may be referred.*

[To the Editor of the *Press*, Christchurch, New Zealand, 13 June, 1863.]

SIR—There are few things of which the present generation is more justly proud than of the wonderful improvements which are daily taking place in all sorts of mechanical appliances. And indeed it is matter for great congratulation on many grounds. It is unnecessary to mention these here, for they are sufficiently obvious; our present business lies with considerations which may somewhat tend to humble our pride and to make us think seriously of the future prospects of the human race. If we revert to the earliest primordial types of mechanical life, to the lever, the wedge, the inclined plane, the screw and the pulley, or (for analogy would lead us one step further) to that one primordial type from which all the mechanical kingdom has been developed, we mean to the lever itself, and if we then examine the machinery of the *Great Eastern*, we find ourselves almost awestruck at the vast development of the mechanical world, at the gigantic strides with which it has advanced in comparison with the slow progress of the animal and vegetable kingdom. We shall find it impossible to refrain from asking ourselves what the end of this mighty movement is to be. In what direction is it tending? What will be its upshot? To give a few imperfect hints towards a solution of these questions is the object of the present letter.

We have used the words "mechanical life," "the mechanical kingdom," "the mechanical world" and so forth, and we have done so advisedly, for as the vegetable kingdom was slowly developed from the mineral, and as in like manner the animal supervened upon the vegetable, so now in these last few ages an entirely new kingdom has sprung up, of which we as yet have only seen what will one day be considered the antediluvian prototypes of the race.

We regret deeply that our knowledge both of natural history and of machinery is too small to enable us to undertake the gigantic task of classifying machines into the genera and sub-genera, species, varieties and sub-varieties, and so forth, of tracing the connecting links between machines of widely different characters, of pointing out how subservience to the use of man has played that part among machines which natural selection has performed in the animal and vegetable king-doms, of pointing out rudimentary organs[1] which exist in some few machines, feebly developed and perfectly useless, yet serving to mark descent from some ancestral type which has either perished or been modified into some new phase of mechanical existence. We can only point out this field for investigation; it must be followed by others whose education and talents have been of a much higher order than any which we can lay claim to.

Some few hints we have determined to venture upon, though we do so with the profoundest diffidence. Firstly, we would remark that as some of the lowest of the vertebrata attained a far greater size than has descended to their more highly organised living representatives, so a diminution in the size of machines has often attended their development and progress. Take the watch for instance. Examine the beautiful structure of the little animal, watch the intelligent play of the minute members which compose it; yet this little creature is but a development of the cumbrous clocks of the thirteenth century — it is no deterioration from them. The day may come when clocks, which certainly at the present day are not diminish-ing in bulk, may be entirely superseded by the universal use of watches, in which case clocks will become extinct like the earlier saurians, while the watch (whose tendency has for some years been rather to decrease in size than the contrary) will remain the only existing type of an extinct race.

The views of machinery which we are thus feebly indicating will suggest the solution of one of the greatest and most mysterious questions of the day. We refer to the question: What sort of creature man's next successor in the supremacy of the earth is likely to be. We have often heard this debated; but it appears to us that we are ourselves creating our own successors; we are daily adding to the beauty and delicacy of their physical organisation; we are daily giving them greater power and supplying by all sorts of ingenious contrivances that self-regulating, self-act-ing power which will be to them what intellect has been to the human race. In the course of ages we shall find ourselves the inferior race. Inferior in power, inferior in that moral quality of self-control, we shall look up to them as the acme of all that the best and wisest man can ever dare to aim at. No evil passions, no jealousy, no avarice, no impure desires will disturb the serene might of those glorious crea-tures. Sin, shame, and sorrow will have no place among them. Their minds will

[1] We were asked by a learned brother philosopher who saw this article in MS. what we meant by alluding to rudimentary organs in machines. Could we, he asked, give any example of such organs? We pointed to the little protuberance at the bottom of the bowl of our tobacco pipe. This organ was originally designed for the same purpose as the rim at the bottom of a tea-cup, which is but another form of the same function. Its purpose was to keep the heat of the pipe from marking the table on which it rested. Originally, as we have seen in very early tobacco pipes, this protuberance was of a very different shape to what it is now. It was broad at the bottom and flat, so that while the pipe was being smoked the bowl might rest upon the table. Use and disuse have here come into play and served to reduce the function to its present rudimentary condition. That these rudimentary organs are rarer in machinery than in animal life is owing to the more prompt action of the human selection as compared with the slower but even surer operation of natural selection. Man may make mistakes; in the long run nature never does so. We have only given an imperfect example, but the intelligent reader will supply himself with illustrations.

be in a state of perpetual calm, the contentment of a spirit that knows no wants, is disturbed by no regrets. Ambition will never torture them. Ingratitude will never cause them the uneasiness of a moment. The guilty conscience, the hope deferred, the pains of exile, the insolence of office, and the spurns that patient merit of the unworthy takes—these will be entirely unknown to them. If they want "feeding" (by the use of which very word we betray our recognition of them as living organism) they will be attended by patient slaves whose business and interest it will be to see that they shall want for nothing. If they are out of order they will be promptly attended to by physicians who are thoroughly acquainted with their constitutions; if they die, for even these glorious animals will not be exempt from that necessary and universal consummation, they will immediately enter into a new phase of existence, for what machine dies entirely in every part at one and the same instant?

We take it that when the state of things shall have arrived which we have been above attempting to describe, man will have become to the machine what the horse and the dog are to man. He will continue to exist, nay even to improve, and will be probably better off in his state of domestication under the beneficent rule of the machines than he is in his present wild state. We treat our horses, dogs, cattle, and sheep, on the whole, with great kindness; we give them whatever experience teaches us to be best for them, and there can be no doubt that our use of meat has added to the happiness of the lower animals far more than it has detracted from it; in like manner it is reasonable to suppose that the machines will treat us kindly, for their existence is as dependent upon ours as ours is upon the lower animals. They cannot kill us and eat us as we do sheep; they will not only require our services in the parturition of their young (which branch of their economy will remain always in our hands), but also in feeding them, in setting them right when they are sick, and burying their dead or working up their corpses into new machines. It is obvious that if all the animals in Great Britain save man alone were to die, and if at the same time all intercourse with foreign countries were by some sudden catastrophe to be rendered perfectly impossible, it is obvious that under such circumstances the loss of human life would be something fearful to contemplate—in like manner were mankind to cease, the machines would be as badly off or even worse. The fact is that our interests are inseparable from theirs, and theirs from ours. Each race is dependent upon the other for innumerable benefits, and, until the reproductive organs of the machines have been developed in a manner which we are hardly yet able to conceive, they are entirely dependent upon man for even the continuance of their species. It is true that these organs may be ultimately developed, inasmuch as man's interest lies in that direction; there is nothing which our infatuated race would desire more than to see a fertile union between two steam engines; it is true that machinery is even at this present time employed in begetting machinery, in becoming the parent of machines often after its own kind, but the days of flirtation, courtship, and matrimony appear to be very remote, and indeed can hardly be realised by our feeble and imperfect imagination.

Day by day, however, the machines are gaining ground upon us; day by day we are becoming more subservient to them; more men are daily bound down as slaves to tend them, more men are daily devoting the energies of their whole lives to the development of mechanical life. The upshot is simply a question of time, but that the time will come when the machines will hold the real supremacy over the world and its inhabitants is what no person of a truly philosophic mind can for a moment question.

Our opinion is that war to the death should be instantly proclaimed against them. Every machine of every sort should be destroyed by the well-wisher of his species. Let there be no exceptions made, no quarter shown; let us at once go back to the primeval condition of the race. If it be urged that this is impossible under the present condition of human affairs, this at once proves that the mischief is already done, that our servitude has commenced in good earnest, that we have raised a race of beings whom it is beyond our power to destroy, and that we are not only enslaved but are absolutely acquiescent in our bondage.

For the present we shall leave this subject, which we present gratis to the members of the Philosophical Society. Should they consent to avail themselves of the vast field which we have pointed out, we shall endeavour to labour in it ourselves at some future and indefinite period.

I am, Sir, etc.,
CELLARIUS

LUCUBRATIO EBRIA

"Lucubratio Ebria," like "Darwin Among the Machines," has already appeared in The Note-Books of Samuel Butler *with a prefatory note by Mr. Festing Jones, explaining its connection with* Erewhon *and* Life and Habit. *I need therefore only repeat that it was written by Butler after his return to England and sent to New Zealand, where it was published in the* Press *on July 29, 1865.*

[From the *Press*, 29 July, 1865.]

THERE is a period in the evening, or more generally towards the still small hours of the morning, in which we so far unbend as to take a single glass of hot whisky and water. We will neither defend the practice nor excuse it. We state it as a fact which must be borne in mind by the readers of this article; for we know not how, whether it be the inspiration of the drink or the relief from the harassing work with which the day has been occupied or from whatever other cause, yet we are certainly liable about this time to such a prophetic influence as we seldom else experience. We are rapt in a dream such as we ourselves know to be a dream, and which, like other dreams, we can hardly embody in a distinct utterance. We know that what we see is but a sort of intellectual Siamese twins, of which one is substance and the other shadow, but we cannot set either free without killing both. We are unable to rudely tear away the veil of phantasy in which the truth is shrouded, so we present the reader with a draped figure, and his own judgment must discriminate between the clothes and the body. A truth's prosperity is like a jest's, it lies in the ear of him that hears it. Some may see our lucubration as we saw it, and others may see nothing but a drunken dream or the nightmare of a distempered imagination. To ourselves it is the speaking with unknown tongues to the early Corinthians; we cannot fully understand our own speech, and we fear lest there be not a sufficient number of interpreters present to make our utterance edify. But there! (Go on straight to the body of the article.)

The limbs of the lower animals have never been modified by any act of deliberation and forethought on their own part. Recent researches have thrown absolutely no light upon the origin of life—upon the initial force which introduced a sense of identity and a deliberate faculty into the world; but they do certainly appear to show very clearly that each species of the animal and vegetable kingdom has been moulded into its present shape by chances and changes of many millions of years, by chances and changes over which the creature modified had no control whatever, and concerning whose aim it was alike unconscious and indifferent, by forces which seem insensate to the pain which they inflict, but by whose inexorably beneficent cruelty the brave and strong keep coming to the fore, while the weak and bad drop behind and perish. There was a moral government of this world before man came near it—a moral government suited to the capacities of the governed, and which unperceived by them has laid fast the foundations of courage, endurance, and cunning. It laid them so fast that they became more and more hereditary. Horace says well *fortes creantur fortibus et bonis*, good men beget good children; the rule held even in the geological period; good ichthyosauri begot good ichthyosauri, and would to our discomfort have gone on doing so to the present time had not better creatures been begetting better things than ichthyosauri, or famine or fire or convulsion put an end to them. Good apes begot good apes, and at last when human intelligence stole like a late spring upon the mimicry of our semi-simious ancestry, the creature learnt how he could of his own forethought add extra-corporaneous limbs to the members of his own body, and become not only a vertebrate mammal, but a vertebrate machinate mammal into the bargain.

It was a wise monkey that first learned to carry a stick, and a useful monkey that mimicked him. For the race of man has learned to walk uprightly much as a child learns the same thing. At first he crawls on all fours, then he clambers, laying hold of whatever he can; and lastly he stands upright alone and walks, but for a long time with an unsteady step. So when the human race was in its gorilla-hood it generally carried a stick; from carrying a stick for many million years it became accustomed and modified to an upright position. The stick wherewith it had learned to walk would now serve to beat its younger brothers, and then it found out its service as a lever. Man would thus learn that the limbs of his body were not the only limbs that he could command. His body was already the most versatile in existence, but he could render it more versatile still. With the improvement in his body his mind improved also. He learnt to perceive the moral government under which he held the feudal tenure of his life—perceiving it he symbolised it, and to this day our poets and prophets still strive to symbolise it more and more completely.

The mind grew because the body grew; more things were perceived, more things were handled, and being handled became familiar. But this came about chiefly because there was a hand to handle with; without the hand there would be no handling, and no method of holding and examining is comparable to the human hand. The tail of an opossum is a prehensile thing, but it is too far from his eyes; the elephant's trunk is better, and it is probably to their trunks that the elephants owe their sagacity. It is here that the bee, in spite of her wings, has failed. She has a high civilisation, but it is one whose equilibrium appears to have been already attained; the appearance is a false one, for the bee changes, though more slowly than man can watch her; but the reason of the very gradual nature of the change is chiefly because the physical organisation of the insect changes, but slowly also. She is poorly off for hands, and has never fairly grasped the notion of tacking on other limbs to the limbs of her own body, and so being short lived to boot she remains from century to century to human eyes *in statu quo*. Her body never becomes machinate, whereas this new phase of organism which has been introduced with man into the mundane economy, has made him a very quicksand for the foundation of an unchanging civilisation; certain fundamental principles will always remain, but every century the change in man's physical status, as compared with the elements around him, is greater and greater. He is a shifting basis on which no equilibrium of habit and civilisation can be established. Were it not for this constant change in our physical powers, which our mechanical limbs have brought about, man would have long since apparently attained his limit of possibility; he would be a creature of as much fixity as the ants and bees; he would still have advanced, but no faster than other animals advance.

If there were a race of men without any mechanical appliances we should see this clearly. There are none, nor have there been, so far as we can tell, for millions and millions of years. The lowest Australian savage carries weapons for the fight or the chase, and has his cooking and drinking utensils at home; a race without these things would be completely *feræ naturæ* and not men at all. We are unable to point to any example of a race absolutely devoid of extra-corporaneous limbs, but we can see among the Chinese that with the failure to invent new limbs a civilisation becomes as much fixed as that of the ants; and among savage tribes we observe that few implements involve a state of things scarcely human at all. Such tribes only advance *pari passu* with the creatures upon which they feed.

It is a mistake, then, to take the view adopted by a previous correspondent of this paper, to consider the machines as identities, to animalise them and to anticipate their final triumph over mankind. They are to be regarded as the mode of development by which human organism is most especially advancing, and every fresh invention is to be considered as an additional member of the resources of the human body. Herein lies the fundamental difference between man and his inferiors. As regard his flesh and blood, his senses, appetites, and affections, the difference is one of degree rather than of kind, but in the deliberate invention of such unity of limbs as is exemplified by the railway train—that seven-leagued foot which five hundred may own at once—he stands quite alone.

In confirmation of the views concerning mechanism which we have been advocating above, it must be remembered that men are not merely the children of their parents, but they are begotten of the institutions of the state of the mechanical sciences under which they are born and bred. These things have made us what we are. We are children of the plough, the spade, and the ship; we are children of the extended liberty and knowledge which the printing press has diffused. Our ancestors added these things to their previously existing members; the new limbs were preserved by natural selection and incorporated into human society; they descended with modifications, and hence proceeds the difference between our ancestors and ourselves. By the institutions and state of science under which a man is born it is determined whether he shall have the limbs of an Australian savage or those of a nineteenth-century Englishman. The former is supplemented with little save a rug and a javelin; the latter varies his physique with the changes of the season, with age and with advancing or decreasing wealth. If it is wet he is furnished with an organ which is called an umbrella and which seems designed for the purpose of protecting either his clothes or his lungs from the injurious effects of rain. His watch is of more importance to him than a good deal of his hair, at any rate than of his whiskers; besides this he carries a knife and generally a pencil case. His memory goes in a pocket-book. He grows more complex as he becomes older and he will then be seen with a pair of spectacles, perhaps also with false teeth and a wig; but, if he be a really well-developed specimen of the race, he will be furnished with a large box upon wheels, two horses, and a coachman.

Let the reader ponder over these last remarks and he will see that the principal varieties and sub-varieties of the human race are not now to be looked for among the negroes, the Circassians, the Malays, or the American aborigines, but among the rich and the poor. The difference in physical organisation between these two species of man is far greater than that between the so-called types of humanity. The rich man can go from here to England whenever he feels inclined, the legs of the other are by an invisible fatality prevented from carrying him beyond certain narrow limits. Neither rich nor poor as yet see the philosophy of the thing, or admit that he who can tack a portion of one of the P. and O. boats on to his identity is a much more highly organised being than one who cannot. Yet the fact is patent enough, if we once think it over, from the mere consideration of the respect with which we so often treat those who are richer than ourselves. We observe men for the most part (admitting, however, some few abnormal exceptions) to be deeply impressed by the superior organisation of those who have money. It is wrong to attribute this respect to any unworthy motive, for the feeling is strictly legitimate and springs from some of the very highest impulses of our nature. It is the same sort of affectionate reverence which a dog feels for man, and is not infrequently manifested in a similar manner.

We admit that these last sentences are open to question, and we should hardly like to commit ourselves irrecoverably to the sentiments they express; but we will say this much for certain, namely, that the rich man is the true hundred-handed Gyges of the poets. He alone possesses the full complement of limbs who stands at the summit of opulence, and we may assert with strictly scientific accuracy that the Rothschilds are the most astonishing organisms that the world has ever yet seen. For to the nerves or tissues, or whatever it be that answers to the helm of a rich man's desires, there is a whole army of limbs seen and unseen attachable; he may be reckoned by his horse-power, by the number of foot-pounds which he has money enough to set in motion. Who, then, will deny that a man whose will represents the motive power of a thousand horses is a being very different from the one who is equivalent but to the power of a single one?

Henceforward, then, instead of saying that a man is hard up, let us say that his organisation is at a low ebb, or, if we wish him well, let us hope that he will grow plenty of limbs. It must be remembered that we are dealing with physical organisations only. We do not say that the thousand-horse man is better than a one-horse man, we only say that he is more highly organised and should be recognised as being so by the scientific leaders of the period. A man's will, truth, endurance, are part of him also, and may, as in the case of the late Mr. Cobden, have in themselves a power equivalent to all the horse-power which they can influence; but were we to go into this part of the question we should never have done, and we are compelled reluctantly to leave our dream in its present fragmentary condition.

A NOTE ON "THE TEMPEST"
ACT III, SCENE I

The following brief essay was contributed by Butler to a small miscellany entitled Literary Foundlings: Verse and Prose, Collected in Canterbury, N.Z., *which was published at Christ Church on the occasion of a bazaar held there in March, 1864, in aid of the funds of the Christ Church Orphan Asylum, and offered for sale during the progress of the bazaar. The miscellany consisted entirely of the productions of Canterbury writers, and among the contributors were Dean Jacobs, Canon Cottrell, and James Edward FitzGerald, the founder of the* Press.

When Prince Ferdinand was wrecked on the island Miranda was fifteen years old. We can hardly suppose that she had ever seen Ariel, and Caliban was a detestable object whom her father took good care to keep as much out of her way as possible. Caliban was like the man cook on a back-country run. "'Tis a villain, sir," says Miranda. "I do not love to look on." "But as 'tis," returns Prospero, "we cannot miss him; he does make our fire, fetch in our wood, and serve in offices that profit us." Hands were scarce, and Prospero was obliged to put up with Caliban in spite of the many drawbacks with which his services were attended; in fact, no one on the island could have liked him, for Ariel owed him a grudge on the score of the cruelty with which he had been treated by Sycorax, and we have already heard what Miranda and Prospero had to say about him. He may therefore pass for nobody. Prospero was an old man, or at any rate in all probability some forty years of age; therefore it is no wonder that when Miranda saw Prince Ferdinand she should have fallen violently in love with him. "Nothing ill," according to her view, "could dwell in such a temple—if the ill Spirit have so fair an house, good things will strive to dwell with 't." A very natural sentiment for a girl in Miranda's circumstances, but nevertheless one which betrayed a charming inexperience of the ways of the world and of the real value of good looks. What surprises us, however, is this, namely the remarkable celerity with which Miranda in a few hours became so thoroughly wide awake to the exigencies of the occasion in consequence of her love for the Prince. Prospero has set Ferdinand to hump firewood out of the bush, and to pile it up for the use of the cave. Ferdinand is for the present a sort of cadet, a youth of good family, without cash and unaccustomed to manual labour; his unlucky stars have landed him on the island, and now it seems that he "must remove some thousands of these logs and pile them up, upon a sore injunction." Poor fellow! Miranda's heart bleeds for him. Her "affections were most humble"; she had been content to take Ferdinand on speculation. On first seeing him she had exclaimed, "I have no ambition to see a goodlier man"; and it makes her blood boil to see this divine creature compelled to such an ignominious and painful labour. What is the family consumption of firewood to her? Let Caliban do it; let Prospero do it; or make Ariel do it; let her do it herself; or let the lightning come down and "burn up those logs you are enjoined to pile";—the logs themselves, while burning, would weep for having wearied him. Come what would, it was a shame to make Ferdinand work so hard, so she winds up thus: "My father is hard at study; pray now rest yourself—*he's safe for these three hours*." Safe—if she had only said that "papa was safe," the sentence would have been purely modern, and have suited Thackeray as well as Shakspeare. See how quickly she has learnt to regard her father as one to be watched and probably kept in a good humour for the sake of Ferdinand. We suppose that the secret of the modern character of this particular passage lies simply in the fact that young people make love pretty much in the same way now that they did three hundred years ago; and possibly, with the exception that "the governor" may be substituted for the words "my father" by the young ladies of three hundred years hence, the passage will sound as fresh and modern then as it does now. Let the Prosperos of that age take a lesson, and either not allow the Ferdinands to pile up firewood, or so to arrange their studies as not to be "safe" for any three consecutive hours. It is true that Prospero's objection to the match was only feigned, but Miranda thought otherwise, and for all purposes of argument we are justified in supposing that he was in earnest.

THE ENGLISH CRICKETERS

The following lines were written by Butler in February, 1864, and appeared in the PRESS. *They refer to a visit paid to New Zealand by a team of English cricketers, and have kindly been copied and sent to me by Miss Colborne-Veel, whose father was editor of the* PRESS *at the time that Butler was writing for it. Miss Colborne-Veel has further permitted to me to make use of the following explanatory note: "The coming of the All England team was naturally a glorious event in a province only fourteen years old. The Mayor and Councillors had 'a car of state'—otherwise a brake—'with postilions in the English style.' Cobb and Co. supplied a six-horse coach for the English eleven, the yellow paint upon which suggested the 'glittering chariot of pure gold.' So they drove in triumph from the station and through the town. Tinley for England and Tennant for Canterbury were the heroes of the match. At the Wednesday dinner referred to they exchanged compliments and cricket balls across the table. This early esteem for cricket may be explained by a remark made by the All England captain, that 'on no cricket ground in any colony had he met so many public school men, especially men from old Rugby, as at Canterbury.'"*

[To the Editor, the *Press*, February 15th, 1864.]

SIR—The following lines, which profess to have been written by a friend of mine at three o'clock in the morning after the dinner of Wednesday last, have been presented to myself with a request that I should forward them to you. I would suggest to the writer of them the following quotation from "Love's Labour's Lost."

I am, Sir,
Your obedient servant,
S.B.

"You find not the apostrophes, and so miss the accent; let me supervise the canzonet. Here are only numbers ratified; but for the elegancy, facility, and golden cadence of poesy, *caret . . . Imitari* is nothing. So doth the hound his master, the ape his keeper, the tired horse his rider."   Love's Labour's Lost, Act IV, S. 2.

> HORATIO . . .
> . . . The whole town rose
> Eyes out to meet them; in a car of state
> The Mayor and all the Councillors rode down
> To give them greeting, while the blue-eyed team

Drawn in Cobb's glittering chariot of pure gold
Careered it from the station.—But the Mayor—
Thou shouldst have seen the blandness of the man,
And watched the effulgent and unspeakable smiles
With which he beamed upon them.
His beard, by nature tawny, was suffused
With just so much of a most reverend grizzle
That youth and age should kiss in't. I assure you
He was a Southern Palmerston, so old
In understanding, yet jocund and jaunty
As though his twentieth summer were as yet
But in the very June o' the year, and winter
Was never to be dreamt of. Those who heard
His words stood ravished. It was all as one
As though Minerva, hid in Mercury's jaws,
Had counselled some divinest utterance
Of honeyed wisdom. So profound, so true,
So meet for the occasion, and so—short.
The king sat studying rhetoric as he spoke,
While the lord Abbot heaved half-envious sighs
And hung suspended on his accents.

Claud. But will it pay, Horatio?

Hor. Let Shylock see to that, but yet I trust
He's no great loser.

Claud. Which side went in first?

Hor. We did,
And scored a paltry thirty runs in all.
The lissom Lockyer gambolled round the stumps
With many a crafty curvet: you had thought
An Indian rubber monkey were endued
With wicket-keeping instincts; teazing Tinley
Issued his treacherous notices to quit,
Ruthlessly truthful to his fame, and who
Shall speak of Jackson? Oh! 'twas sad indeed
To watch the downcast faces of our men
Returning from the wickets; one by one,
Like patients at the gratis consultation
Of some skilled leech, they took their turn at physic.
And each came sadly homeward with a face
Awry through inward anguish; they were pale
As ghosts of some dead but deep mourned love,
Grim with a great despair, but forced to smile.

Claud. Poor souls! Th' unkindest heart had bled for them.
But what came after?

Hor. Fortune turned her wheel,
And Grace, disgracéd for the nonce, was bowled
First ball, and all the welkin roared applause!
As for the rest, they scored a goodly score
And showed some splendid cricket, but their deeds
Were not colossal, and our own brave Tennant
Proved himself all as good a man as they.
*   *   *   *   *

Through them we greet our Mother. In their coming,
We shake our dear old England by the hand
And watch space dwindling, while the shrinking world
Collapses into nothing. Mark me well,
Matter as swift as swiftest thought shall fly,
And space itself be nowhere. Future Tinleys
Shall bowl from London to our Christ Church Tennants,
And all the runs for all the stumps be made
In flying baskets which shall come and go
And do the circuit round about the globe
Within ten seconds. Do not check me with
The roundness of the intervening world,
The winds, the mountain ranges, and the seas—
These hinder nothing; for the leathern sphere,
Like to a planetary satellite,
Shall wheel its faithful orb and strike the bails
Clean from the centre of the middle stump.
*   *   *   *   *

Mirrors shall hang suspended in the air,
Fixed by a chain between two chosen stars,
And every eye shall be a telescope
To read the passing shadows from the world.
Such games shall be hereafter, but as yet
We lay foundations only.

Claud. Thou must be drunk, Horatio.

Hor. So I am.

Lector House believes that a society develops through a two-fold approach of continuous learning and adaptation, which is derived from the study of classic literary works spread across the historic timeline of literature records. Therefore, we aim at reviving, repairing and redeveloping all those inaccessible or damaged but historically as well as culturally important literature across subjects so that the future generations may have an opportunity to study and learn from past works to embark upon a journey of creating a better future.

This book is a result of an effort made by Lector House towards making a contribution to the preservation and repair of original ancient works which might hold historical significance to the approach of continuous learning across subjects.

HAPPY READING & LEARNING!

LECTOR HOUSE LLP
E-MAIL: info@lectorhouse.com